Nyedja Fialho Morais Barbosa

Kernel Smoothing of Rainfall Data in the Northeast

Nyedja Fialho Morais Barbosa

Kernel Smoothing of Rainfall Data in the Northeast

Interpolation of more than 26 million rainfall data points in Northeast Brazil

ScienciaScripts

Imprint
Any brand names and product names mentioned in this book are subject to trademark, brand or patent protection and are trademarks or registered trademarks of their respective holders. The use of brand names, product names, common names, trade names, product descriptions etc. even without a particular marking in this work is in no way to be construed to mean that such names may be regarded as unrestricted in respect of trademark and brand protection legislation and could thus be used by anyone.

Cover image: www.ingimage.com

This book is a translation from the original published under ISBN 978-620-2-17644-6.

Publisher:
Sciencia Scripts
is a trademark of
Dodo Books Indian Ocean Ltd. and OmniScriptum S.R.L publishing group

120 High Road, East Finchley, London, N2 9ED, United Kingdom
Str. Armeneasca 28/1, office 1, Chisinau MD-2012, Republic of Moldova, Europe
Printed at: see last page
ISBN: 978-620-7-41034-7

Table of contents:

I dedicate this work to the memory of my father, Pedro, and my aunt and uncle, Teca and Morais.

Thanks

First of all, I thank God for allowing me to get this far, for his great love and mercy poured out on my life, without him I would be nothing. To Him be all the honor and glory now and forever!

I would like to thank my mother, Salete, my brothers Érika and Júnior, and my father Pedro, who is already in God's arms. I'm sure they were the greatest supporters of my success, always supporting me, teaching me, loving me...

I would especially like to thank my husband, Luiz André, who supported me so much in this achievement. His love, understanding and affection have strengthened me and pushed me to achieve this victory.

I would like to thank my aunts Penha and Teca for the love and support they gave me and my sister Érika. I'm sure they were angels sent by God to help us through this difficult time in our lives.

I can't forget to thank my friends, companions and warriors, who have helped me so much to get this far... My special thanks go to Érika, Priscila, Renata, Neto, Marystella, Clara (and Sophia), Danila, Samuel and Silvio.

I would like to thank all the teachers and staff at Biometria, especially Professors Borko and Tatijana, Professor Eufrázio, and the secretary Marco.

Finally, I would like to thank FACEPE, which helped me a lot by funding my studies. Without its support, I would not have been able to realize this dream.

Summary

The Northeast of Brazil has great climatic diversity and is considered a very complex region, arousing the interest of scholars from all over the world. The rainfall regime in this region is considered to be seasonal because it behaves more intensely in three internal zones of the region, in different periods of the year, lasting three months, as well as being strongly influenced by the incidence of El Niño, La Niña and other phenomena acting on the basins of the Pacific and Atlantic Oceans. In this work, the mathematical-computational technique of Kernel Smoothing interpolation was applied to rainfall data for the Northeast Region of Brazil collected from 1904 to 1998, from 2,283 conventional weather stations located in all the states of the Northeast. The calculations were carried out on the "Cluster Neumann" GPU of the Postgraduate Program in Biometrics and Applied Statistics of the Department of Statistics and Informatics at UFRPE using the "Kernel" software written in C and Cuda. This tool made it possible to interpolate more than 26 million rainfall measurements over the whole of the Northeast, making it possible to generate rainfall intensity maps over the whole region, as well as making estimates in areas where data was missing, and calculating statistics for the Northeast's rainfall in general and seasonal terms. According to the interpolations carried out, it was possible to detect the driest and wettest years during the period studied, the spatial distribution of rainfall in each month, as well as the characteristics of rainfall in times of El Niño and La Niña.

Keywords: Interpolation; Kernel Smoothing; Rainfall in the Northeast.

Chapter 1

1 Introduction

The northeastern region of Brazil is known for its great climatological diversity. Its rainfall regime has been studied by scientists all over the world in an attempt to explain the behavior of this phenomenon, which is so fickle and highly influenced by temperature anomaly disturbances in the Pacific Ocean.

Phenomena such as El Niño and La Niña generally affect the rate of rainfall, causing times of great drought and times of great rainfall. In addition, the Northeast contains semi-arid regions with a concentration of rainfall in a short period of the year, and long dry spells, resulting in various effects that make life difficult for the Northeastern population.

There is a lot of rainfall data for the Northeast, but organizing, understanding and visualizing it, which is fundamental for modelling this phenomenon, is still a major challenge for planning the analysis and use of water resources.

Generally, precipitation data is stored from conventional and automatic weather stations and geostationary and polar-orbiting weather satellites. The information collected from weather stations has a long time series, but is not able to cover all areas of the globe, while the data obtained from satellite images covers the entire surface of the earth, but does not have a long time series.

There are several studies in the world's scientific literature, such as those by Li and Shao (2010) and Garcia-Pintado et al. (2009), aimed at taking advantage of the benefits and reducing the shortcomings of data sources, with the general objective of establishing standards and rules for processing data, extracting the maximum amount of information contained therein, as well as establishing procedures and rules for future measurements.

Although the mathematical/computational technique *Parzen-Rosenblatt Window Method* (ROSENBLATT, 1956), (PARZEN, 1962), (EPANECHNIKOV, 1969) was proposed several decades ago, the computational challenge for its practical implementation in the current case has only become possible with recent revolutionary advances in parallel processing hardware, in general the GPGPU'S, *General Purpose Graphics Processing Units,* and in particular with the use of the *"Neumann"* GPU *cluster* installed with resources from the *"Casadinho"* project on the premises of the Postgraduate Program in Biometrics and Applied Statistics of the Department of Statistics and Informatics at UFRPE.

The aim of this work is to use the Kernel Smoothing technique (ROSENBLATT, 1956), (PARZEN, 1962), (EPANECHNIKOV, 1969), to construct a continuous estimate of the spatio-temporal density of rainfall for the Northeast region, using rainfall data from the former SUDENE () which is managed by the Pernambuco Water and Climate Agency, APAC (), corresponding to a mass of more than 26 million observations from 2.283 conventional weather stations throughout the Northeast, collected between 1904 and 1998.

This work will investigate the characteristics of rainfall over the Northeast throughout the period studied, as well as in specific periods where there are records of the incidence of the El Niño and La Niña phenomena (TRENBERTH, 1997), as well as phases of the Pacific Decadal Oscillation (PAULA, 2009). The spatial behavior of rainfall over the Northeast in each month of the year will also be analyzed, as well as in times of seasonal rainfall (MENEGHETTI; FERREIRA, 2009), (RAO et al., 1993).

Mapping will make it possible to provide an overview of rainfall distribution on a spatial and temporal scale, thus establishing a basis for future studies of rainfall dynamics.

Knowledge of the spatial distribution of rainfall in the Northeast will be of the utmost importance in helping the competent public bodies to draw up plans and measures to improve the use of the region's water resources and to set targets to benefit the places most affected by the dry seasons.

Chapter 2

2 Literature review

Since the beginning, man has tried to understand and unravel the mysteries of nature, so he began to observe the behaviour of natural phenomena and looked for ways to organize these observations so that he could, in a way, prepare for the future.

According to Mendonca and Danni-Oliveira (2007), the Egyptians, observing the regime of floods and ebbs in the River Nile, were led to reflect on how the climatic elements were formed and related them to the fertility of the soils in the river's floodplain. However, the Egyptians were not the only ones who observed the phenomena of nature. There are records of other peoples, such as the Greeks and Romans.

All of these past contributions have been important in enabling humanity, through the observation of natural phenomena, to understand the world around them, take advantage of the benefits of nature and look for ways to avoid natural disasters.

The observations of these phenomena have contributed to a better understanding of atmospheric and regional dynamics, helping to advance climate research.

2.1   Noyóes de Meteorología e Climatología

Meteorology, by definition, is the study of the atmosphere and its laws with the aim of predicting weather variations, and climatology is the science that describes climates, explains them and classifies them by zones, and both go hand in hand to better explain natural phenomena (COSTA, 2006).

According to Mendonca and Danni-Oliveira (2007), the meteorologist's duty is to apply the laws of physics and mathematical techniques to observed phenomena, while the climatologist uses statistical techniques on the data in order to make inferences about climate information. It is important to note that climatologists base their studies on existing meteorology, establishing a partnership between the study of climate and weather.

In meteorology, a distinction is made between weather and climate. Weather is the state of the atmosphere at a given time and place or the state of the atmosphere in relation to its effects on human life and activities. Climate, on the other hand, is the synthesis of the weather in a given place for a given period of time, according to (VIEIRA; PICULLI, 2009).

According to Hartmann (1994), the climate of the Earth's surface is very flexible, ranging from very hot regions to extremely cold places, from the intense drought of the desert to the humidity of tropical forests.

In order to understand the elements and factors that make up climatology, a brief survey of definitions will follow.

2.1.1 Climatic elements

Climatic elements are meteorological quantities that transmit their peculiar properties and characteristics to the environment, influencing a given region. According to Costa (2006) and Silva (2006), the main climatic elements are presented below:

- **Temperature :**

The definitions of temperature are related to the physiological sensation in the human body, but these concepts are quite imprecise. Temperature indicates the thermal state of the body and heat is the manifestation of energy that can be transformed into work and passed from one body to another when their temperatures are different.

- **Air humidity :**

Most of the Earth's surface is covered by water. The upper part of the water circulating over the planet is continually evaporating. The evaporated water remains in the atmosphere until it returns to the earth in the form of rain, snow, dew, etc. The presence of water in the atmosphere is called humidity.

- **Precipitation :**

Precipitation occurs when a cloud can no longer hold the excess moisture it contains and, due to gravitational force, the drops of water fall, thus renewing the hydrological cycle. The best known forms of precipitation are rain, drizzle, snow and others.

- **Wind :**

Wind is the horizontal movement of air that occurs when two different regions have different atmospheric pressures due to temperature variations. Winds are characterized by their strength and the direction in which they are blowing. According to their variations in speed and length of time, winds are called gusts, breezes, storms and hurricanes, among others.

- **Cloudiness :**

Nebulosity is the fraction of the celestial vault that is surrounded by clouds that are very close together.

- **Atmospheric Pressure :**

The pressure of atmospheric air is the result of the force exerted in all directions by the weight of the air. As a result of the various constant movements of the air, variations in its temperature and water vapor content, the weight of the atmospheric air at a given point varies constantly. Pressure, therefore, like temperature, never stabilizes.

2.1.2 Climatic factors

Climatic elements are directly influenced by a number of climatic factors that act on the entire globe. The main factors that influence the climate are presented below, according to Costa (2006) and Silva (2006):

- **Latitude :**

The earth is supposedly divided into geographical coordinates formed by imaginary lines. By convention, the lines parallel to the equatorial plane, which are called parallels, and the meridians, which are imaginary lines parallel to the Greenwich meridian, connect the north and south poles. Latitude, then, is the distance measured in degrees from a given point on the planet to the equator.

- **Longitude :**

Longitude is the angle between the meridian plane of any location on the Earth's surface and the Greenwich meridian plane. Longitude is counted from the Greenwich meridian eastwards (E) and westwards (O) up to 180° .

- **Altitude :**

Altitude is the vertical distance, measured in meters, from the equatorial plane, which has sea level as a reference, to the maximum height of the object measured, and is considered the third geographical coordinate for locating a body on planet earth.

- **Air Mass :**

An air mass is a large amount of the atmosphere that has the same physical properties, such as temperature and humidity. When these masses move over a certain region, they lose their original characteristics and acquire new ones due to exchanges with the new surface and the vertical movements that begin to take place inside them, so that the thermodynamic properties come into equilibrium.

- **Continentality :**

Continentality is the influence caused by the ocean and is expressed by the distance from the earth's surface to the sea.

- **Sea currents :**

Ocean currents are masses of water that circulate along oceans and seas, maintaining their own characteristics of color, temperature and salinity, without acquiring characteristics of the places they are traveling through. It is believed that the formation of these currents is the result of the influence of winds and the earth's rotational movement.

- **Relief :**

Relief is the set of forms of the earth's surface, from the bottom of the oceans to the surfaces with the highest altitudes. The main types of relief are mountains, plateaus, plains and depressions. Relief is the result of the action of external forces (such as rain, ice, the seas, animals and plants, as well as the action of man himself), and internal forces (tectonism, volcanism and seismic tremors), which have acted on the planet over the years, modifying its forms in various ways.

- **Vegetation :**

Vegetation is the set of plants specific to a given region that grow naturally without human interference. Vegetation is directly linked to climate, hydrography, relief and soil. It regulates the flow of the water, carbon and nitrogen cycles, and affects the characteristics of the soil in terms of its volume, chemistry, texture and fertility conditions in a given region.

2.2   Rainfall

In nature there are many phenomena related to water, such as dew, frost, rain, drizzle, snow, among others. All these forms of water are known as hydrometeors (SILVA, 2006).

In the formation of rain, water droplets evaporated from lakes, rivers and oceans rise in the form of water vapor and form clouds. When a cloud cannot hold the excess moisture it contains, the water droplets fall to earth in the form of rain, thus renewing the water cycle (COSTA, 2006).

The amount of rainfall in a given region during any given period is measured using an instrument called a rain gauge, which gives the measurement in millimetres, i.e. the amount of liters of water that have fallen per square metre of the earth's surface projection, and is characterized by its duration and intensity (COSTA,

2006).

2.3   Rainfall data

For the analysis of meteorological phenomena, a unified observation system is needed, which allows the atmosphere to be explored at both surface and upper levels in extremely short time intervals so that phenomena can be monitored in the best possible way.

The World Meteorological Organization, WMO, is the specialized body for coordinating meteorological activities on an operational basis, maintained by the United Nations since 1950. This organization develops the World Meteorological Surveillance program, VMM, maintaining the exchange of meteorological information between various countries (VIEIRA; PICULLI, 2009).

According to Vieira and Piculli (2009), the program developed by the WMO consists of three systems:

1. **World Observation System**: Around 10,000 ground stations, mostly on the continents and in the northern hemisphere, 7,000 merchant ships, 3,000 commercial aircraft, automatic platforms, satellites and radars, focused on the quality and quantity of observations.

2. **World Data Preparation System**: Consisting of National Meteorological Centers (CMN), Regional Meteorological Centers (CMR) and World Meteorological Centers (CMM- Washington, Moscow and Melbourne), aimed at processing data and preparing forecasts;

3. **World** Telecommunications **System**: With national telecommunications centers (CNT).

There are two ways of obtaining meteorological precipitation data. They can be collected from weather stations or weather satellites.

2.3.1   Weather station data

Meteorological stations are basic units for capturing meteorological data on the earth. They are generally placed in places where there is no human or natural interference, so that there is no doubt as to the reliability of their data (COSTA, 2006).

Surface observations are systematized and standardized by the **World Meteorological Organization, WMO**, and are controlled from the maintenance of equipment, methods of obtaining results and data analysis. In this way, the information on the meteorological parameters observed can be compared with other observations and can also characterize the instantaneous state of the atmosphere (VIEIRA; PI- CULLI, 2009).

According to Vieira and Piculli (2009), there are two types of surface weather stations: conventional weather stations and automatic weather stations. Conventional Weather Stations require the presence of an observer to collect and organize the data, while Automatic Weather Stations use sensors that operate using electronic signals, which allows the data to be stored more quickly and reliably.

2.3.2   Meteorological Satellite Data

Meteorological satellites can be geostationary or polar-orbiting, each with a specific route and resolution.

Polar-orbiting meteorological satellites cover the Earth's circumference from pole to pole, positioned between 800 and 1,200 km from the Earth's surface, providing images in nominal ranges of approximately 3,000 km. Geostationary satellites, on the other hand, are positioned in equatorial orbit, remaining in a fixed position in relation to the observed point, located at a height of approximately 35,800 km, providing images 24 hours a day (FERREIRA, 2006).

When it comes to monitoring the weather, data from geostationary satellites is better than that from polar-orbiting satellites, as it is updated continuously, making it possible to monitor a given region with greater fidelity.

Data from weather satellites is transmitted by the ViSSR sensor, "*Visible and Infrared Spin Scan Radiometer*", and sent by the platform directly to ground stations. The data is processed in the control center and sent back to the satellites for retransmission at a slower rate, and can be accessed by several users. Another way of distributing the data is through the WEFAX system, "*Weather Fascimile*", with which the data is broadcast by radio according to schedules established by the WMO (FERREIRA, 2006).

2.3.2.1   Reanalysis data

The reanalysis process takes place in two stages: Firstly, the data captured by satellite is analyzed and interpolated by the system with the help of models used to predict the weather over the region of interest. Then, based on the data collected, new data is interpolated in areas where it is not possible to capture any information about the phenomenon (FREITAS et al., 2010).

According to Kalnay et al. (1996), in the reanalysis process the data is produced and reanalyzed four times a day and is also summarized daily and monthly, based on Universal Time (UTC).

One of the best-known reanalysis projects is ERA-40, which gathers data from September 1957 to August 2002, produced by the European Center for Medium-Range Weather Forecasts (ECMWF) among other institutions (UPPALA et al., 2005).

Reanalysis data is very convenient for filling in the gaps left in some areas where information is lacking. To analyze collected data, pre-filling existing gaps in precipitation time series, a convenient way is to use reanalysis data from the NCEP (BERNARDO; MOLION, 2002).

2.4 Characteristics of the Brazilian Northeast

The Northeastern Region of Brazil is located between 1° and 18° 30' South latitude and 34° 30' and 40° 20' West longitude of Greenwich, with a territorial extension of approximately 1,540,827 Km² , with a relief composed of low coastal plains, low valleys below 500m, between surfaces that are at elevations of 800m, in the Serra da Borborema region, and 1,200m in the Chapada da Diamantina (NIMER, 1989).

Rainfall and radiation are considered to be the main meteorological variables in the Northeast. With a non-uniform rainfall regime, there is inter-annual and seasonal variation which affects the amount of rainfall in the region. As such, the period of rainfall modulated by the trade winds can cause major damage, or influence the growth or development of agricultural crops, influencing the local economy (MENEGHETTI; FERREIRA, 2009).

The Northeast differs from other Brazilian regions in its complexity. The characteristic vegetation of the Northeast is Atlantic Forest, Cocais Forest, Cerrado, Caatinga, Coastal Vegetation and Riparian Forests. Its main basins are: São Paulo Basin

Francisco Basin, Parnaíba Basin, East Northeast Atlantic Basin, West Northeast Atlantic Basin and East Atlantic Basin. The region has four types of climate: humid equatorial, humid coastal, tropical and semi-arid, the latter being the most widespread.

2.4.1　The Northeastern Semi-Arid

A region is considered semi-arid when there is a deficiency and/or irregularity of rainfall. In this region, evaporation exceeds precipitation, and there are usually long periods of intense drought (CARITAS-BRASILEIRA, 2001).

This region has strong sunshine, high temperatures and a low volume of rainfall. The concentration of rain in a short period of time, ranging from three to four months on average, is a dominant regional characteristic, which is why the population suffers from long periods of drought.

According to (SA; SILVA, 2010), water demand generally exceeds supply in these regions and one of the biggest problems is the combination of irregular rainfall and high temperatures, causing high rates of water deficiency.

The main characteristics of semi-arid areas are: Average annual rainfall equal to or less than 800 mm; Average sunshine of 2,800 h/year; Rainfall regime marked by irregularity (space/time); Dominance of the Caatinga Ecosystem (diversity) and Rainfall limitations with low soil retention (SUDENE, 2012).

A major concern in the semi-arid region, apart from the long dry periods, is the process of desertification. This phenomenon is worsening as a result of scarce resources and the inter-annual variability of rainfall (CONTI, 2005).

Figure 1 shows the division of the semi-arid region between the states (IBGE, 2012).

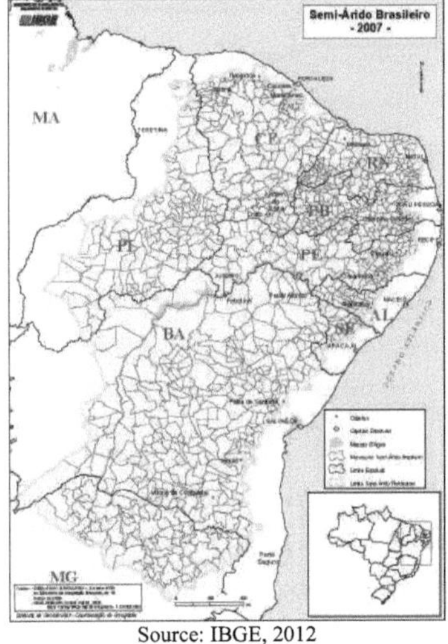

Source: IBGE, 2012

Figure 1: Brazilian semi-arid region according to IBGE estimates (2007)

2.4.2 Rain in the Northeast

The Northeast has different rainfall regimes in specific areas. In the north, the rainy season generally falls between March and May, in the south and southeast, it is common for rain to fall between December and February, and in the east of the region, rain generally falls between May and July. The spatial variation of annual rainfall over this region is quite high, as annual rainfall varies from 300 mm in the driest region to 2,000 mm on the east coast, according to studies by (RAO et al., 1993), (MENEGHETTI; FERREIRA, 2009).

The region is considered anomalous in the tropical continents because, in contrast to other regions in this latitudinal range, it has a semi-arid climate, which is due to the relatively low rainfall values over most of the region, according to (NOBRE et al., 1986). The atmospheric circulation over the region is strongly modulated and modified by the thermodynamic patterns over the Pacific and Atlantic Ocean basins (FERREIRA; MELLO, 2005), influencing the behavior of rainfall over the region (MOLION; BERNARDO, 2002).

The mechanisms that determine the complexity of the region's rainfall regime, according to (UVO; BERNDTSSON, 1996) are: El Nmo-Southern Oscillation Events (ENOS); Sea Surface Temperature (SST) in the Atlantic Ocean basin, Trade Winds, Sea Level Pressure (SLP); Intertropical Convergence Zone (ITCZ) over the Atlantic Ocean and Cold Fronts and High Level Cyclonic Vortices (HLCV).

In years when there are positive or negative SST anomalies in the basins of these oceans, the Hadley cell, which acts in a meridional direction, and the Walker cell, which acts in a zonal direction, are disturbed, causing strong anomalies in the atmospheric circulation over the tropics, since these cells are displaced from their climatological positions, consequently the intensity and duration of the rainy season in this region are also affected (FERREIRA; MELLO, 2005).

Tropical Atlantic SST variability is the dominant force behind precipitation anomalies in Northeast Brazil, while the remote influence of the Pacific can reinforce these anomalies at certain times, but at other times it can have the opposite effect, so that these anomalies are weakened (UVO et al., 1998 apud ANDREOLI; KAYANO, 2007).

For (NOBRE et al., 1986), it is likely that the most important factor in determining how abundant or deficient the rainfall of a given rainy season will be in the dry part of the Northeast is the latitudinal position of the ITCZ over the Equatorial Atlantic, more specifically over the western portion near the coast of South America. In dry years, the ITCZ usually doesn't cross the equator on its seasonal migration southwards, so it doesn't reach the Northeast.

In rainy years, the North Atlantic Subtropical Anticyclone (Azores High) is more intense than normal, the northeasterly trade winds from the North Atlantic are also more intense and the ITCZ is more prominent to the south. For drier years, the reverse occurs: the South Atlantic anticyclone is more intense, the Azores high is weaker, the South Atlantic trade winds are more intense and the ITCZ is further north than in its normal position, according to (NOBRE et al., 1986).

Historically, the Northeast has always been affected by major droughts or floods. Reports of droughts in the region can be traced back to the 17th century, when the Portuguese arrived in the region. Statistically, 18 to 20 years of drought occur every 100 years (MARENGO, 2007).

2.4.3 Effects of El Niño and La Niña

According to Freire et al. (2011), El Niño and La Niña are large-scale meteorological phenomena characterized by Sea Surface Temperature Anomalies (SSTAs) in the Pacific Ocean, which occur simultaneously with anomalies in the Southern Oscillation Index (SOI). These phenomena affect atmospheric circulation, mainly determining anomalies in rainfall in various regions of the globe.

El Niño and La Niña events tend to alternate every 3.2 years. However, from one event to the next, the interval can vary from 1 to 10 years. The intensity of the events varies greatly from case to case. The most intense El Niño, according to ATSM measurements, occurred in 1982-83 and 1997-98 (COSTA, 2009).

El Niño is characterized by an overheating of the waters of the Equatorial Pacific Ocean. This abnormal warming can affect the local climate, changing the direction of the winds and causing major droughts in the semi-arid region of Brazil. This phenomenon usually lasts from 2 to 7 years, with an average duration of 12 to 18 months.

La Niña is the opposite. The waters of the Western Pacific cool down. This phenomenon causes the South Pacific Subtropical High to strengthen, transporting cold surface waters to the Equatorial Pacific. The consequences are major floods in the Northeast. Their average duration is between nine and twelve months.

According to Trenberth (1997), the periods of El Niño and La Niña effects on Brazil between 1950 and 1998 are described in Tables 1 and 2.

The anomaly that occurs in sea surface temperature is the subject of study for many researchers (CHU, 1983), (HASTENRATH; GREISCHAR, 1993), (HASTENRATH, 1990) and (GRIMM et al., 1998).

In a particular case for investigating this phenomenon in one of the northeastern states, SILVA et al. (2011) stated that the influence of the distribution of sea surface temperature anomalies on rainfall in northeastern Pernambuco is directly related to areas of the Southern Tropical Atlantic and areas of the Equatorial Pacific. Warm waters in the South Atlantic Basin have a positive impact on rainfall, while colder waters have a negative impact on the rainy season in this region of the state. In relation to the Pacific region, when the water temperature is below average there is an increase in rainfall in the north-eastern sector of the state.

Table 1: Periods of El Niño activity between 1950 and 1998

Home	**End**	**Duration in months**	**Intense**
August 1951	February 1952	7	Weak
March 1953	November 1953	9	Weak
April 1957	June 1958	15	Weak
June 1963	February 1964	9	Weak
May 1965	June 1966	14	Moderate
September 1968	March 1970	19	Weak
April 1972	March 1973	12	Moderate
August 1976	March 1977	8	Moderate
July 1977	January 1978	7	Weak
October 1979	April 1980	7	Weak

April 1982	July 1983	16	Strong
August 1986	February 1988	19	Strong
March 1991	July 1992	17	Moderate
February 1993	September 1993	8	Moderate
June 1994	March 1995	10	Moderate
April 1997	April 1998	13	Moderate

Source: Trenberth (1997)

Table 2: La Niña periods between 1950 and 1998

Home	**End**	**Duration in months**	**Intensity**
Margo from 1950	February 1951	12	Strong
June 1954	March 1956	22	Strong
May 1956	November 1956	7	Strong
May 1964	January 1965	9	Moderate
July 1970	January 1972	19	Moderate
June 1973	June 1974	13	Strong
September 1974	April 1976	20	Strong

September 1984	June 1985	10	Strong
May 1988	June 1989	14	Strong
September 1995	March 1996	7	Strong

Source: Trenberth (1997)

According to Marengo (2007), not all periods of drought are related to El Niño. In the last century, for example, it was observed that the years in which the two events coincided were 1900, 1902, 1907, 1915, 1919, 1932-33, 1936, 194144, 1951, 1953, 1958, 1970, 1979-80, 1981, 1982-83, 1986-87, 1991-1992, 1997-1998.

2.4.4 Pacific Decadal Oscillation

The rainfall regime over the Northeast region is also influenced by the Pacific Decadal Oscillation. This phenomenon is characterized by a disturbance in sea surface temperature over the Pacific Ocean basin recently discovered by researchers at the University of Washington in 1996, who discovered a decadal pattern of work done on the variation of fish populations in the North Pacific (MANTUA et al., 1997 apud PAULA, 2009).

In their studies, they realized that the surface temperatures of the Pacific Ocean (TSM) present a configuration with longer-term variations, similar to El Niño, called the Pacific Decadal Oscillation (PDO) and described by (MANTUA et al., 1997 apud MOLION, 2003).

These events generally last 20 to 30 years and come in two phases. The cold phase is characterized by negative SST anomalies in the Tropical Pacific and, at the same time, positive SST anomalies in the Extratropical Pacific in both hemispheres. The warm phase has the opposite configuration, with positive SST anomalies in the Tropical Pacific and negative ones in the Extratropical Pacific (MOLION, 2003).

In the 20th century, there were two hot phases between 1925-1946 and 19771998 and two cold phases between 1910-1924 and 1947-1976 (PAULA, 2009).

During the positive phase of the LTO, there tends to be a greater number of more intense El Niño episodes. Fewer and less intense La Niña episodes. During the negative phase of the LTO, there is a greater occurrence of La Niña episodes, which tend to be more intense, and a lower frequency of El Niños, which tend to be short and rapid (ANDREOLI; KAYANO, 2004 apud SILVA; GALVíNCIO, 2011).

2.5 Interpolation of spatio-temporal data

Historically, in order to reduce the computational cost of estimating densities, a functional or parametric form has been imposed on the data. This parametric form can be quite subjective, however, in most situations, its imposition simplifies the problem considerably, since all that remains is to estimate the parameters using the data sample (FERREIRA, 2007).

In most methods it is necessary to know the probability distribution to which the data belongs, but in some cases it is very difficult to know the parameters of the distribution. In these cases, it is necessary to look for other tools to analyze the behavior of the data.

In climatology, the most commonly used interpolation methods for modeling phenomena are: Weighted inverse distance, Kriging, Minimum curvature, Nearest neighbor, Polynomial regression, Radial basis functions, Shepard's method, Triangulation with linear interpolation, according to (NIKOLOVA; VASSILEV, 2006).

It is important to stress that the more uniformly the data is distributed over the entire study area, or if the data is concentrated in certain clusters of points, very good estimates will be obtained, regardless of the interpolation algorithm used.

2.5.1   Use of Interpolation Methods in Pluviometry

Precipitation is a climatic element that changes according to the different factors that act on each location, causing variations over time and space.

In order to solve some questions about the behavior of rainfall data in a given region, it is necessary to resort to statistical and computational methods, which, by analyzing the database, allow us to analyze the phenomena related to rainfall.

Generally, rainfall data is collected from conventional or automatic weather stations or weather satellites and contains information on the amount (in millimeters) of daily rainwater, the geographical location of the station (altitude. latitude and longitude), among others, latitude and longitude), among others, making it possible to study the variation in the amount of rain that falls, both in relation to a given location on the globe in terms of spatial distribution, and in relation to temporal distribution.

The most widely used methods for modeling rainfall-related phenomena are Kriging and Inverse Weighted Distance. In some cases, the first method is preferred (MATKAN et al., 2010), (OLIVEIRA et al., 2012) and (VERWORN; HA- BERLANDT, 2011). In other cases, the second method is more appropriate (TOMCZAK, 1998).

Many researchers around the world compare these two methods to analyze which one better explains the behavior of their variables, as can be seen in the works developed by (REIS et al., 2005), (SILVA et al., 2011), (HARTKAMP et al., 1999), (MELLO et al., 2003) and (SOENARIO et al., 2010).

The Kernel interpolation technique is also used in rainfall, but infrequently, because it is computationally intensive. The literature on this application in Climatology and Meteorology is still very scarce and has been increasing in recent scientific work (STECK, 2002), (XIONG et al., 2006), (JOU et al., 2012) and (ALI, 1998).

Chapter 3

3 Materials and Methods

3.1 Rainfall data

The rainfall data used in this work comes from observations co-let by the Northeast Development Superintendence, SUDENE, which is under the care of the Pernambuco Water and Climate Agency, APAC.

The database consists of rainfall measurements taken from 2,283 conventional weather stations located in all the states of the Northeast Region, between the years 1904 to 1998, corresponding to 26,496,444 observations.

The spatial distribution of the meteorological stations is not uniform throughout the region studied, but from the interpolations made, it will be possible to estimate the behavior of rainfall throughout the Northeast.

3.2 Methodology

3.2.1 Kernel Smoothing

A very effective method for interpolating point data using non-parametric statistics, studied by (ROSENBLATT, 1956), (PARZEN, 1962), is the Kernel Estimator, or simply Kernel Estimator. This new, computationally intensive technique has recently been used more frequently because of the growth in computing power (PIMENTEL, 2010), (FREITAS et al., 2002), (DUONG, 2007), (PINTO, 2003), (KAWAMOTO, 2012), (CHUNG, 2006), (FERREIRA, 2007).

The Kernel method is widely used to map and estimate the distribution of points in space, using non-parametric statistics through the Kernel function. This interpolation method is based on the primitive histogram method, where no knowledge of the probability distribution of the data is required, only the data is organized into classes according to the frequency with which it appears in the sample. This method can be used to smooth the data and thus estimate an unknown density.

According to Ali (1998), Kernel estimators are weighted moving averages of a target function, where the weight is prescribed through a kernel function that is usually chosen to be a symmetric probability density function with finite variance, which plays the role of a weighting function.

Estimation using the kernel method smoothes the surface, calculating the density of each "window" (or "grid") through interpolation, without modifying the characteristics and variability of the data set, making it possible to estimate values for areas where there were no observations.

The function of the Kernel is to interpolate an intensity value for each cell in a grid, considering a symmetrical function, centered on the cell, using points located up to a certain distance from the center of the cell for the calculation (CARNEIRO; SANTOS, 2003).

Suppose there is a random variable from which a random sample x_1, ..., x_n has been obtained, whose observed values are independent and identically distributed. The kernel estimator, $f_h(x)$, for this sample is given by the mathematical expression given by 3.1:

$$\hat{f}_h(x) = \frac{1}{nh} \sum_{i=1}^{n} K\left(\frac{x - X_i}{h}\right) \qquad (3.1)$$

where n represents the sample size, K(.) represents the Kernel function chosen for the interpolation, and h, the radius of the estimation, also called the smoothing parameter. These parameters are the only ones needed to calculate the estimates (SCHEID, 2004).

Many *probability density functions* are used in this non-parametric method. Table 3 shows the most commonly used ones.

It is important to mention that the Kernel density estimate has the same properties as the chosen kernel function, so it is interesting to choose a smooth and clearly unimodal Kernel symmetrical around zero, says SCOTT (1950).

The choice of density function for the estimation is important, but much more important is the choice of radius, because the radius defines the neighborhood of the point to be interpolated and controls how smooth the density estimation will be.

Table 3: Some Kernel Fungi

	$K(x)$
Uniform Kernel Function	$K(x) = \frac{1}{2}$
Epanechnikov Biweight	$K(x) = \frac{3}{4}(1 - \frac{1}{5}x^2)/\sqrt{5}$
Guassian	$K(x) = \frac{15}{16}(1 - x^2)^2$
Triangular	$K(x) = \frac{1}{\sqrt{2\pi}}e^{-(1/2)x^2}$
	$K(x) = 1 - |x|$

According to Cortes (2004), the smoothing parameter h defines a kind of controller between the bias and the variance of the estimate. If h is too low, you may be failing to smooth the function efficiently (under-smoothing), while you may be over-smoothing the function (over-smoothing) if h is too high.

When each point in the space to be estimated has a value associated with it, the following formula is used to find the estimates:

$$\hat{Y}(x_0) = \sum_{i=1}^{n} K\left(\frac{x_0 - X_i}{h}\right) y(X_i)$$

which represents the amount of attribute per unit area.

If the objective is to find the average value of the attribute, it is sensible to use the expression given by 3.2, which represents the Smother Kernel of the function:

$$\dot{Y}(x_0) \qquad \frac{\sum_{i=1}^{n} K\left(\frac{x_0 - X_i}{h}\right) y(X_i)}{\sum_{i=1}^{n} K\left(\frac{x_0 - X_i}{h}\right)} \qquad (3.2)$$

3.2.2 Computer implementation

The numerical calculations for this work were implemented on the "Cluster Neumann" GPU of the Postgraduate Program in Biometrics and Applied Statistics (PPGBEA), of the Department of Statistics and Informatics at UFRPE, using the "Kernel" software developed at PPGBEA in C and CUDA languages. This tool made it possible to carry out complex calculations which, if performed by the CPU alone, would have cost a lot of time, or would not have been possible at all, since the data corresponds to approximately 26 million observations, with an average volume of 10^9 pixels.

The "Kernel" software used in this work is prepared to read the database, calculate the average daily rainfall of the weather stations, and through the Kernel Smoothing function, using the Gaussian function, estimate the density curve, which represents rainfall, for the entire map of the Northeast region, filling in the gaps with missing data, and estimating the average daily rainfall over the region, in the desired time frame.

After the "Kernel" has been compiled, an input box appears on the computer screen, as shown in Figure 2.

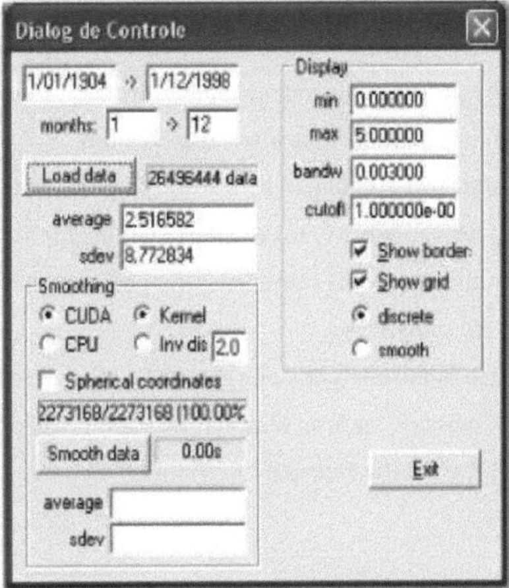

Figure 2 Kernel software input box

The interpolation process is carried out through the following steps:

1 Fill in the blanks for the period in which you want to interpolate (in this case, we used data from 01/01/1904 to 31/12/1998);

2 Select the months that were included in the interpolation (in this case we have the months from January to December);

3 Press the "Load data" key to load the data (you can see the number of observations used in the interpolation);

4 . The "Cuda" and "Kernel" options are selected in the "Smoohhin" window;

5 . click on the "Smoohhdat"" button to start the interpolation (when it is possible to check the estimated time for the end of the interpolation);

6 . Entallaa-ee anntappolapdopdom annfarmapdoda mdd¡ad¡ár¡aVeatcc¡p¡raQdoplvvlal in the rnrlisrdo period, as well as its prdrao deviation, and visurlizraao o mrpr dr estimrtivr rainfall over the entire Northeast Region.

Within the "Displry" jrnelr it is possible to select the maximum and minimum values to be considered, in order to compose the color scale used in the mrpr, and we can still control the smoothing parameter, h. In this case, as we are working with daily rainfall averages, the maximum and minimum values will be 0mm and 5mm, respectively. The color scale will therefore vary from red, when it indicates a low amount of rainfall, to blue, when it indicates a high amount of rainfall.

Also in the "Display" window, there are the options "Show border", to show the borders of the boundaries of each state on the map, and "Show grid", to show the grid that represents the latitude and longitude on the map. (In this way, it is possible to locate the position of each Conventional Weather Station.

Within the "Display" window, you can select one of the buttons: "discrete" or "smooth". The first provides the geographical location of the weather stations that have precipitation information for the period analyzed, and the second generates the smoothed data.

In the "Smoohting" window, there are two initially empty spaces. In the first space, you will see the result of the average daily rainfall for the period analyzed, calculated by the "Kernel" software, and in the second space you will see the standard deviation of the data after interpolation.

The accumulated annual rainfall was calculated by multiplying the annual daily average by the number of days in the year. (In this way, it was possible to make interpolations for all the years in the period analyzed and then find the accumulated rainfall, as well as classifying the years according to the amount of rainfall into dry, very dry, normal, rainy and very rainy years.

(The interpolations were carried out to analyze some specific periods relevant to the Northeast region with regard to the spatial distribution of rainfall, such as seasonal periods in some locations in the region, years of incidence of El Niño, La Niña and the Pacific Decadal Oscillation.

Chapter 4

4 Results and Discussion

The following are the results of interpolations made for rainfall data in the Northeast of Brazil using the kernel smoothing technique on data from SUDENE, from 1904 to 1998.

To begin the study, all the observations in the database were interpolated.

The software used made it possible to locate the position of each weather station that contributed to the research, as shown in Figure 3.

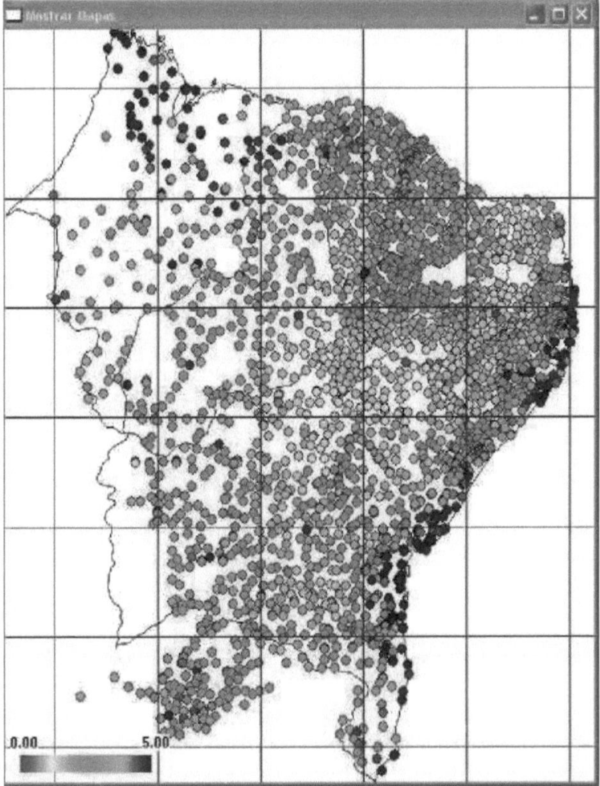

Figure 3 Meteorological Stations in the Northeast Region of Brazil from 1904 to 1998

Figure 3 shows that the spatial distribution of stations over the northeastern area is not homogeneous, with many areas uncovered by information.

To fill in these large areas of missing data, daily rainfall was interpolated using the Kernel Smoothing method. The result can be seen in Figure 4.

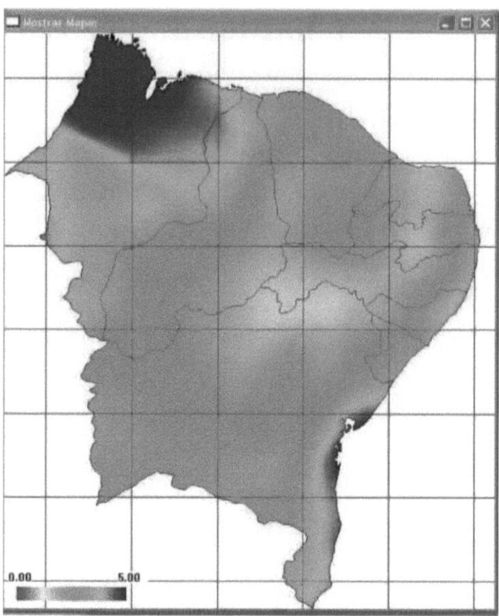

Figure 4 Interpolation of average daily rainfall data in the Northeast of Brazil from 1904 to 1998

Figure 4 shows that there is a large area over the state of Maranhão where the amount of rainfall stands out from the other areas, followed by the east coast of Bahia. And, as expected, there is a low level of rainfall over the semi-arid region of the Northeast.

The average daily rainfall over the Northeast for the period 1904 to 1998, obtained by interpolation, was 2.83 mm with a standard deviation of 0.97 mm.

The amount of rainfall accumulated in each year was also analyzed, except for the years 1904 to 1909, as there was very little information. The calculation was made by multiplying the annual daily average by 365, which corresponds to the number of days in the year.

Table 4 shows the amount of precipitation accumulated annually, in millimeters, from 1910 to 1998, based on the interpolation of observational data.

Table 4: Accumulated Annual Precipitation (mm) of the Northeast Region

Year	Precipitated	Year	Precipitated	Year	Precipitated	Year	Precipitated
1910	648,86	1934	1082,25	1957	1060,29	1980	1049,23
1911	557,57	1935	1138,69	1958	737,26	1981	851,00
1912	1041,16	1936	954,36	1959	781,68	1982	778,33
1913	1041,45	1937	1000,32	1960	1106,02	1983	732,34

1914	976,99	1938	787,23	1961	864,10	1984	987,29
1915	605,46	1939	957,36	1962	898,15	1985	1654,87
1916	1122,26	1940	1192,75	1963	1045,07	1986	1052,04
1917	1208,54	1941	817,71	1964	1351,01	1987	755,40
1918	1071,49	1942	865,82	1965	864,83	1988	1090,80
1920	729,59	1943	866,33	1966	1050,51	1989	1281,62
1921	1129,53	1944	971,92	1967	1182,16	1990	662,15
1922	1157,56	1945	1059,67	1968	1110,51	1991	851,36
1923	957,50	1946	799,64	1969	1058,21	1992	680,10
1924	1680,93	1947	1250,71	1970	913,16	1993	333,24
1925	1043,79	1948	973,35	1971	1007,84	1994	839,03
1926	1299,43	1949	947,61	1972	938,85	1995	984,37
1927	985,68	1950	1038,94	1973	1117,97	1996	684,63
1928	855,45	1951	684,85	1974	1474,86	1997	933,27
1929	1146,03	1952	903,45	1975	1112,78	1998	593,05
1930	903,59	1953	684,85	1976	851,11	-	
1931	865,74	1954	807,45	1977	1135,66	-	
1932	611,78	1955	902,75	1978	1140,77	-	
1933	1052,91	1956	895,13	1979	957,72	-	

The average accumulated annual rainfall for the period studied is 962.9064 mm, with a standard deviation of 222.5017 mm. It can be seen that some years have much lower accumulated rainfall and others much higher than the average. The great oscillation between the amount of rainfall over the Northeast region is due to the incidence of meteorological and climatic phenomena acting on the region.

Figure 5 shows the histogram of accumulated annual rainfall for the Northeast from 1910 to 1998.

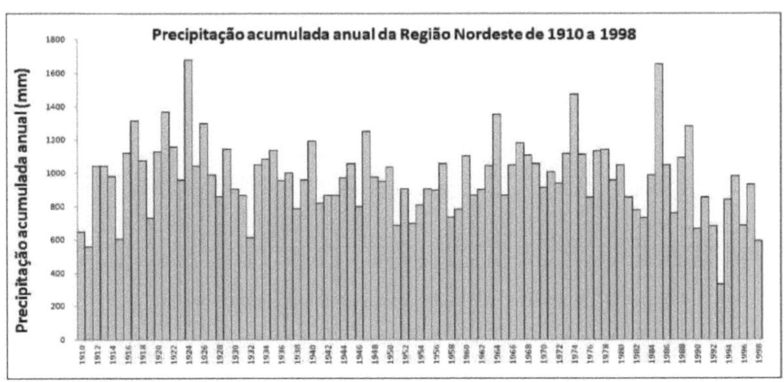

Figure 5 Accumulated Annual Precipitation for the periods 1910 to 1998 in the Northeast

Figure 5 shows that the behavior of the data does not follow any specific pattern over the years.

Based on the accumulated rainfall of each year, it is possible to classify the years as very dry (MS), dry (S), normal (N), rainy (C) and very rainy (MC), grouping the accumulated rainfall into percentiles of 20% each, according to (AL VES; REPELLI, 1992).

The range of accumulated annual rainfall, according to the data analyzed in Table 4, varied between 333.24 mm and 1680.93 mm, which is why the limits chosen were 300 mm and 1700 mm. Thus, a very dry year is one in which the accumulated rainfall was less than 580 mm, a dry year is one in which the accumulated annual rainfall was between 580 mm and 860 mm, a normal year is one in which the accumulated annual rainfall was between 860 mm and 1140 mm, a rainy year is one in which the accumulated annual rainfall was between 1140 mm and 1420 mm, and a very rainy year is one in which the accumulated annual rainfall was more than 1420 mm.

The classification of each year analyzed, according to the classification criteria, is shown in Table 5.

Table 5: Classification of years in relation to Annual Accumulated Precipitation

Accumulated Annual Precipitation	Classification	Years
300mm < *Prec.* < 580mm	Very dry (MS)	1911, 1993
580mm < *Prec.* < 860mm	Dry (S)	1910, 1915, 1919, 1932, 1938, 1941,1946, 1951, 1953, 1954 1958, 1959, 1976, 1981, 1982 1983, 1990, 1991, 1992,1994 1996, 1998
860mm < *Prec.* < 1140mm	Normal (N)	1914, 1918, 1920, 1923, 1925, 1927, 1928, 1930, 1933, 1934, 1935, 1936, 1937, 1939, 1942, 1943, 1944, 1945, 1948, 1949, 1950, 1952, 1955, 1956, 1957, 1960, 1961, 1962, 1963, 1965, 1966, 1968, 1969, 1970, 1971, 1972, 1973, 1975, 1977, 1979, 1980, 1984, 1986, 1987, 1988, 1997
1140mm < *P rec.* < 1420mm	Rainy (C)	1912, 1913, 1916, 1917, 1921, 1922, 1926, 1929, 1940, 1947, 1964, 1967, 1978, 1989
1420mm < *P rec.* < 1700mm	Very rainy (MC)	1924, 1974, 1985

Based on the results in Table 5, we can highlight 1911 and 1993 as the driest years within the observed period and 1924, 1974 and 1985 as having the highest accumulated annual rainfall.

To investigate whether there are significant differences over time, the behavior of the phenomenon can be analyzed in intervals. For this purpose, three intervals were chosen, each with an amplitude of 30 years.

Table 6 shows the daily average of interpolated rainfall data over the Northeast for the intervals chosen.

Table 6: Average daily rainfall in 30-year intervals over Northeast Brazil

Period	Average of interpolated data	Standard Deviation	CV
1910-1938	2,8126	0,9780	34,77
1939-1968	2,7927	0,8233	29,48
1969-1998	2,6617	0,8241	30,96

The results in Table 6 show that the Coefficient of Variation was low in all three periods, so it can be said that the data appears to be homogeneous.

Figure 6 shows the interpolation made for each period.

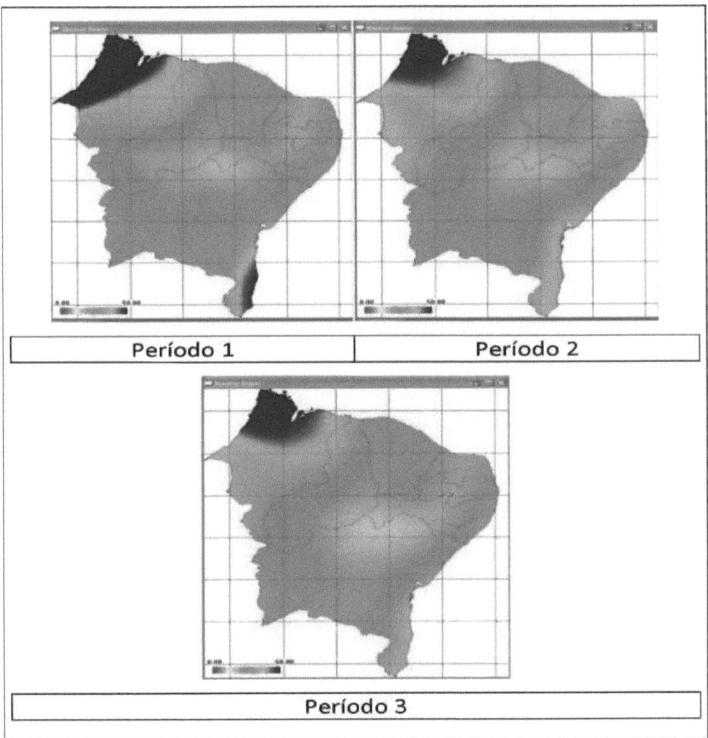

Figure 6 Interpolated average daily rainfall data for the Northeast region of Brazil from 1904 to 1998 at 30-year intervals.

It is possible to see from Table 6 that the amount of rainfall has been decreasing with each period, but looking at Figure 6, one might think that there has been no significant change since the three figures show similar behavior, with a higher concentration of rain in parts of the states of Maranhão and Bahia, and a lower amount over the semi-arid region of the Northeast.

Using the software, it was also possible to calculate the average daily rainfall for individual months, discarding unnecessary months from the period entered by the user in the input box. To carry out this interpolation, the months of interest were entered below the period. This made it possible to calculate the average daily rainfall for each month, adding up all the rainfall in each month, without interference from the other months.

The estimated average daily rainfall for each month during the period from 1904 to 1998 over the Northeast can be seen in Table 7.

Table 7: Average daily monthly rainfall for the period 1904 to 1998 in Northeast Brazil

Month	Average	Standard Deviation	CV
January	4,0190	1,7601	43,79
February	4,9275	2,1994	44,64
March	5,7832	2,6366	45,59
April	4,8373	2,6399	54,57
May	2,5229	2,2476	89,09
June	1,4898	1,6269	109,21
July	1,1583	1,4720	127,08
August	0,6779	0,8942	131,90
September	0,7076	0,6479	91,57
October	1,4008	1,0053	71,77
November	2,7718	1,9117	68,97
December	3,4013	1,7959	52,80

Based on the data provided by the software for the average daily rainfall of each month over the Northeast Region during the entire period analyzed, it is possible to see that the months of January, February, March and April concentrate a large amount of rain, and the months of August, September and October show the least amount of rainfall over the region.

Figure 7 shows the average rainfall for each month for the period 1904 to 1998 throughout the Northeast.

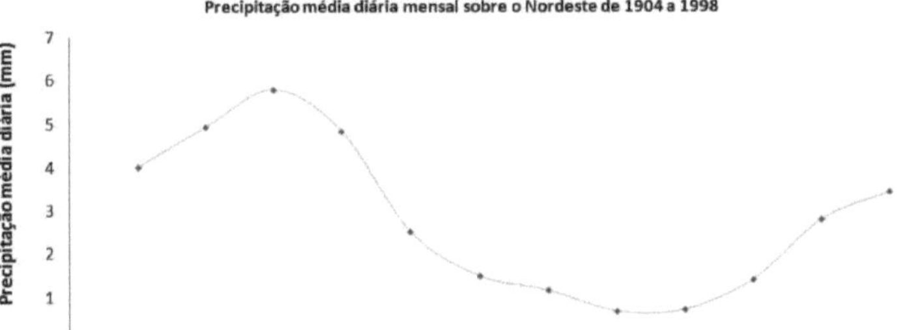

Figure 7 Scatter plot of monthly average daily rainfall over the Northeast from 1904 to 1998

With the help of Figure 7, it can be seen that the average daily rainfall in each month behaves systematically, with phases where there is a greater concentration of rain over the Northeast, and phases where there is less rain.

We could say that from September to March, the rainfall trend over the Northeast is increasing, while from April to August, the rainfall trend over the region is decreasing. The inflection points are in March and August. This behavior may be the result of meteorological phenomena that act on the Northeast of Brazil at specific times of the year.

Figure 8 shows the spatial interpolation of rainfall over the Northeast region from January to December from 1904 to 1998, for each month of the year.

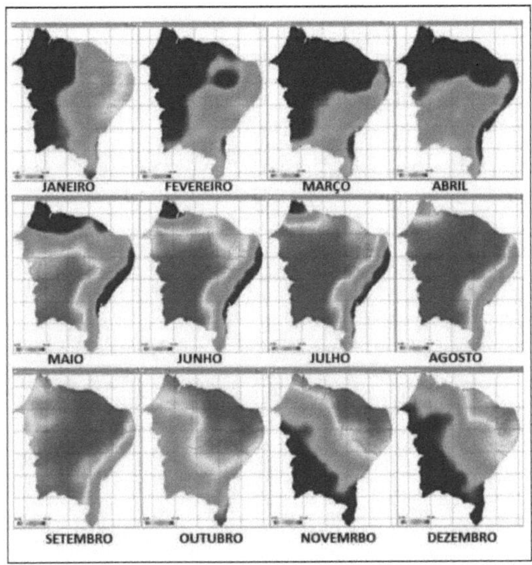

Figure 8 Interpolated monthly rainfall data for the Northeast region of Brazil from 1904 to 1998.

Figure 8 shows that the spatial distribution of rainfall is not uniform across the Northeast, with months where rain is concentrated in specific areas.

According to Meneghetti and Ferreira (2009), the Northeast has three distinct rainfall regimes: in the north, it rains most between March and May, in the south and southeast, it rains from December to February, and in the east, it rains between May and July.

Table 8 shows the amount of rain recorded in each seasonal period, with the exception of the last interval, which does not correspond to any rainy period.

Table 8: Seasonal rainfall for the period 1904 to 1998 in Northeast Brazil

Month	Average	Standard Deviation	CV
December, January, February	4,2528	1,9073	44,85
March, April, May	4,3887	2,3415	53,35
June, July, August	1,1066	1,3066	118,07
September, October, November	1,6064	1,0040	62,50

Among the periods analyzed, we can highlight the first two periods, where the average daily rainfall was higher than the rest of the periods. To visualize the spatial distribution of rainfall during the periods

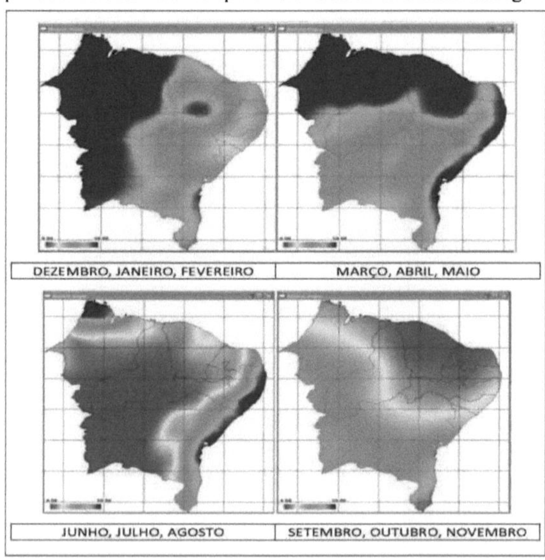

studied, see Figure 9.

Figure 9 Interpolation of average daily rainfall data in the Northeast Region of Brazil from 1904 to 1998 for each local rainy season.

From what can be seen in Figure 9, in each season analyzed over the years studied, there is a greater concentration of rain in certain regions over the months, making the winter weather different in each location.

Rainfall in the Northeast is also influenced by a number of climatic phenomena.

In order to analyze the distribution of rainfall during El Niño and La Niña periods, interpolations were made of the rainfall data recorded during the periods indicated by Trenberth (1997) as being periods of incidence of these phenomena.

Table 9 shows the average daily rainfall, in millimeters, for the El Niño periods, bearing in mind that the periods mentioned correspond to weak, moderate and strong events.

Table 9: Average daily rainfall during El Niño periods in the Northeast

Period	Intensity	Average	Standard Deviation
Aug/1951 to Feb/1952	Weak	1,5734	0,7011
Mar/1953 to Nov/1953	Weak	1,7442	0,7249
Apr/1957 to Jun/1958	Weak	2,8164	0,8965
Jun/1963 to Feb/1964	Weak	3,2215	0,9852
May 1965 to June 1966	Moderate	2,4088	0,7783
Sep/1968 to Mar/1970	Weak	3,1080	0,8856
Apr/1972 to Mar/1973	Moderate	2,7242	0,8872
Aug/1976 to Mar/1977	Moderate	2,9293	0,9824
Jul/1977 to Jan/1978	Weak	2,2235	0,9350
Oct/1979 to Apr/1980	Weak	4,6741	1,6333
Apr/1982 to Jul/1983	Strong	1,8691	0,1297

Period	Intensity	Value 1	Value 2
Aug/1986 to Feb/1988	Strong	2,1284	0,7647
Mar/1991 to Jul/1992	Moderate	2,1857	1,1047
Feb/1993 to Sep/1993	Moderate	0,6052	0,3638
Jun/1994 to Mar/1995	Moderate	2,0045	1,0152
Apr/1997 to Apr/1998	Moderate	2,3961	1,4146

Table 9 shows that the periods from February 1993 to September 1993 and October 1979 to April 1980 had the lowest and highest rainfall recorded, respectively, with 0.8564 mm and 4.6741 mm.

With this result, it is possible to imagine that in some El Niño periods, the average daily amount of rain that falls on the Northeast region is not greatly affected by this phenomenon, or even that some areas are punished by drought and others do not suffer from this problem, while in other periods, such as the period from February 1993 to September 1993, it is imagined that the entire Northeast region suffers.

Figure 10 shows the spatial distribution of rainfall over each period analyzed.

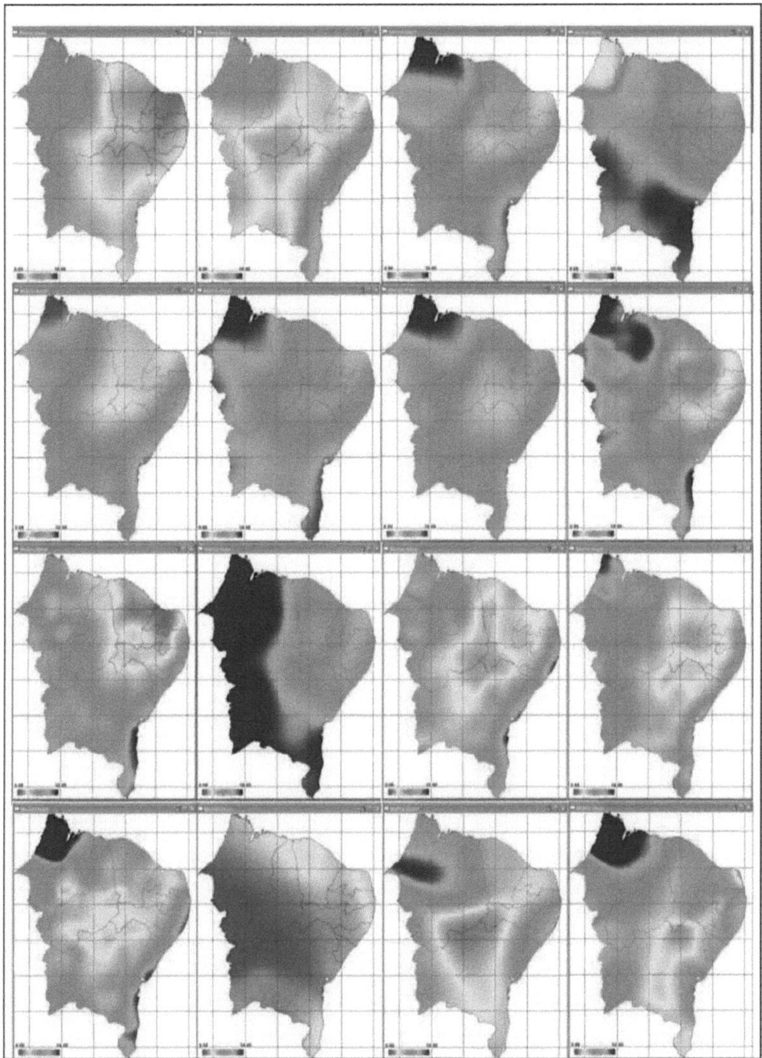

Figure 10: El Niño periods over the Northeast in relation to average daily rainfall

With the help of the spatial interpolation of rainfall data over the Northeast carried out by Kernel Smoothing, it is possible to analyze the spatial distribution of rainfall during El Niño periods, where it can be seen that some periods seem to suffer no influence on their rainfall regime, such as the period from October 1979 to April 1980, and others where some "reddish" areas can be seen on the map, of which we can see that the period from February 1993 to September 1993 showed the greatest rainfall deficit.

Figure 11 shows a graph of the average daily rainfall, in millimeters, during the 16 El Niño periods in the Northeast.

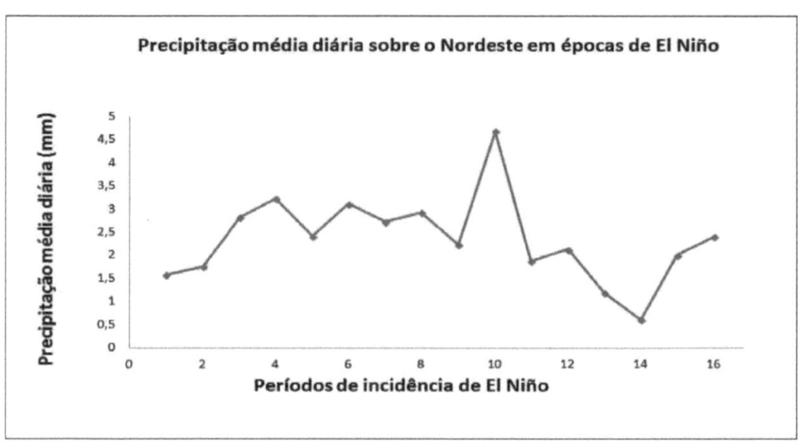

Figure 11: Graph of average daily rainfall during El Niño periods in the Northeast

Figure 11 shows that during El Niño events in the Northeast, the average daily rainfall appears to be similar, except for the isolated cases mentioned above.

It is possible to construct a confidence interval for the average daily rainfall over the Northeast Region in El Niño seasons based on the information extracted from the software used. To do this, we will use the following formula:

$$IC_{(\mu_E)} = \left[\bar{x}_E \pm t_{(\alpha,n-1)} \frac{S_E}{\sqrt{n_E}} \right]$$

where //E is the average daily rainfall during El Niño periods, $\bar{x}_E$ is the estimate of the average daily rainfall during El Niño periods, $t_{(\alpha,n-1)}$ is the quantile of the t distribution (used in this case at a significance level of 5%, with 15 degrees of freedom), S_E is the standard deviation of the average rainfall and n_E is the number of El Niño periods.

The average daily rainfall obtained through the software was 2.4133 mm and the standard deviation was 0.888 mm; the t-distribution quantile used in this calculation was 2.13145; the number of observations corresponded to the number of El Niño events, which in this case was 16.

In this way, we can say with 99% confidence that the average daily rainfall during the El Niño period can belong to the interval [1.7591; 3.0667], based on the information extracted from the sample.

Another phenomenon that also influences the rainfall regime over the Northeast is La Niña, which is why the amount of rainfall collected during its period of action was investigated. Table 10 shows the average daily rainfall during La Niña periods in the Northeast, according to (TRENBERTH, 1997).

Table 10: Average daily rainfall during La Niña periods in the Northeast

Period	Intensity	Average	Standard Deviation
Mar/1950 to Feb/1951	Strong	2,4523	0,9390
Jun/1954 to Mar/1956	Strong	2,2771	0,7252
May/1956 to Nov/1956	Strong	1,4478	0,9653
May/1964 to Jan/1965	Moderate	2,2431	0,9510
Jul/1970 to Jan/1972	Moderate	2,4642	0,8348
Jun/1973 to Jun/1974	Strong	3,9034	1,3409
Sep/1974 to Apr/1976	Strong	2,9706	1,0157
Sep/1984 to Jun/1985	Strong	4,4357	1,6841
May/1988 to June/1989	Strong	2,5857	1,0102

Figure 12 shows the interpolation of precipitation data for the Northeast region during the La Niña period.

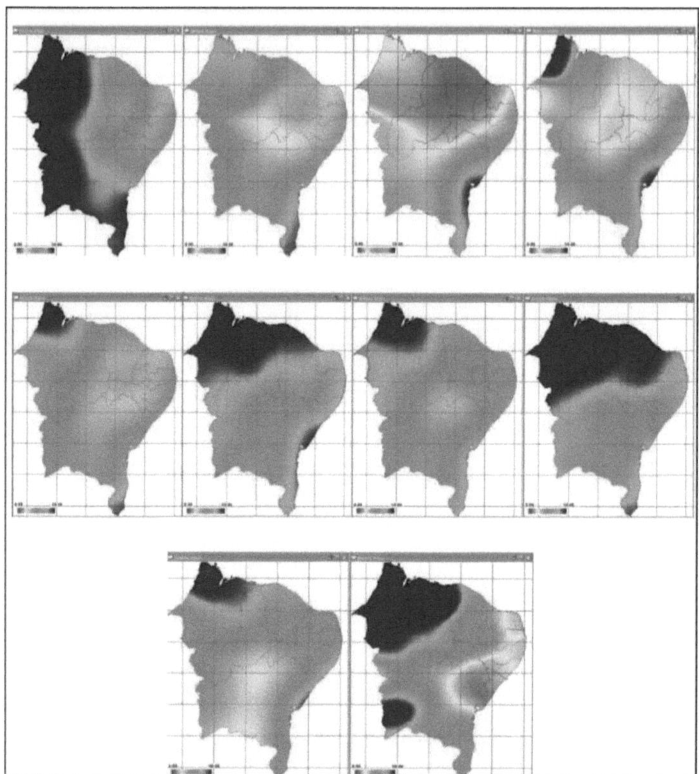

Figure 12: Average daily rainfall during La Niña periods over the Northeast

Figure 12 shows that some periods have little rainfall, such as the period from May 1956 to November 1956, but the majority show plenty of rainy days.

Figure 13 shows the behavior of the daily rainfall averages for each period shown in Table 10.

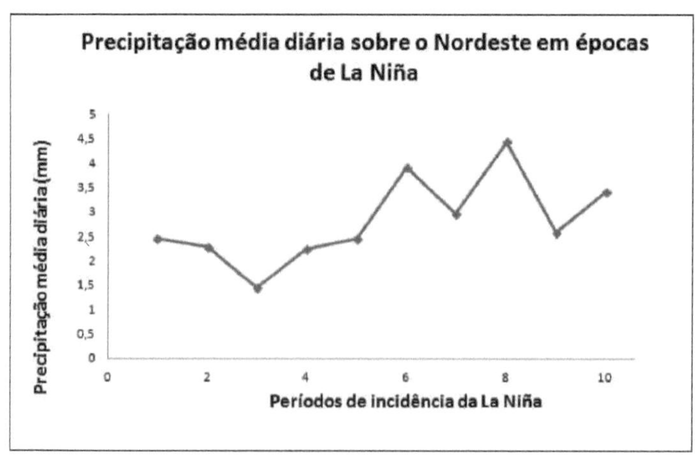

Figure 13 Graph of average daily rainfall during La Niña periods over the Northeast versus average daily rainfall

It seems that the daily rainfall averages have increased since the sixth La Niña period, but it doesn't seem to follow a pattern.

Just as was done to estimate a confidence interval for the average precipitation during El Niño periods, it is possible to construct a confidence interval for the average daily precipitation over the Northeast Region during La Niña periods. To do this, we will use the following formula:

$$IC_{(\mu_L)} \quad \left[\bar{x}_L \pm t_{(\alpha, n-1)} \frac{S_L}{\sqrt{n_L}} \right]$$

where p,L is the average daily rainfall during La Niña periods, $\bar{x}_L$ is the estimate of the average daily rainfall during La Niña periods, $t_{(\alpha-1)}$ is the quantile of the t distribution (used in this case at a significance level of 5%, with 9 degrees of freedom), s_L is the standard deviation of the average rainfall and nL is the number of El Niño periods.

The mean of the average daily rainfall obtained through the software was 2.8821 mm, with a standard deviation of 3.067 mm; the value of the quantile of the t-distribution used in this calculation was 2.262157; the number of observations corresponds to the number of La Niña events, which in this case was 9.

In this way, we can say with 99% confidence that the average daily rainfall during the La Niña period can be contained in the interval [1, 1914; 3, 7273], based on the information extracted from the sample.

It is possible to imagine that during El Niño and La Niña events, the amount of rainfall over the Northeast is different. In times of El Niño, the amount of rainfall over the region is expected to be low, while in times of La Niña, rainfall is expected to be more abundant over the Northeast.

To check whether there is a significant difference between the daily rainfall averages during El Niño and La Niña, a confidence interval will be constructed for the difference in averages, given by the formula:

$$IC_{(\mu_E - \mu_L)} = \left[\bar{x}_E - \bar{x}_L \pm t_{(\alpha, n_E + n_L - 2)} \sqrt{\frac{S_E^2}{n_E} + \frac{S_L^2}{n_L}} \right]$$

where $\mu_E - \mu_L$ is the difference between the mean daily rainfall in El Niño and La Niña periods, respectively; $\bar{x}_E - \bar{x}_L$ is the difference between the estimated mean daily rainfall in El Niño and La Niña periods; $t_{(\alpha, E_{n+L-2})}$ nis the quantile of the t distribution (used in this case at the 5% significance level, with 24 degrees of freedom); S_E^2

S_E^2 are the variances of the El Niño and La Niña periods and nE and nL are the sample sizes.

The interval obtained for the difference in means was [-1, 1444; 0, 3295]. With this result, we cannot say that there is a significant difference between the daily rainfall averages between El Niño and La Niña, because at the 95% confidence level, the interval for the difference in averages, because "zero" is contained in this interval. In other words, we cannot es- tatistically say that it rains less during El Niño than during La Niña. This result may be due to the fact that El Niño and La Niña do not have a homogeneous influence on the whole of the Northeast, since the rainfall regime is seasonal. It is possible that these phenomena influence some areas in isolation.

Another phenomenon that influences the rainfall regime in the Northeast, according to (PAULA, 2009), is the Pacific Decadal Oscillation. The periods of action of this phenomenon, described by (MANTUA et al., 1997 apud MOLION, 2003), were analyzed. The results for each period are shown in Table 11.

Table 11: Precipitation versus Pacific Decadal Oscillation over Northeast Brazil

Phase	Period	Average	Standard Deviation	CV
Cold	1910-1924	3,0001	0,9768	32,55
Hot	1925-1946	2,6952	0,9662	35,85
Cold	1947-1976	2,8669	0,9189	32,05
Hot	1977-1998	2,7292	0,8311	30,45

There doesn't seem to be much difference in the average daily rainfall between the warm and cold phases.

Figure 14 shows the result of interpolating the cold and warm phases of the Decadal Oscillation in relation to the amount of precipitation in the Northeast.

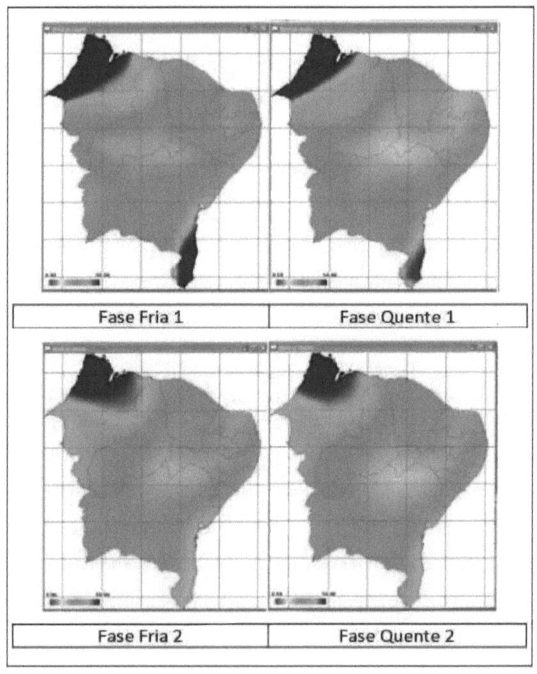

Figure 14 Interpolation of rainfall data in the Northeast Region of Brazil from 1904 to 1998 in the Pacific Decadal Oscillation phase.

Looking at Figure 14, it is possible to see that during the first cold phase and the first hot phase the behavior is roughly similar, with a slight decrease in precipitation over part of Maranhao and Bahia, as well as a small part of the semi-arid region. In the second hot phase and the second cold phase, there is not much difference either. However, the difference between the first cold phase and the second cold phase, as well as from the first hot phase to the second hot phase is clear. It seems that the amount of average daily rainfall has been gradually decreasing.

Conclusions

The Northeast has a great deal of climatological diversity. Rainfall in this region has peculiar characteristics and has been the subject of study by many researchers.

In this work, the spatial and temporal distribution of rainfall over the Northeast was analyzed using rainfall data from the former SUDENE () and managed by the Pernambuco Water and Climate Agency (APAC). The database contained information on the amount of rainfall, in millimeters, collected from 2,283 conventional weather stations, as well as the geographical location of each station, spread across parts of the Northeast, between the years 1904 to 1998, corresponding to more than 26 million observations.

The mathematical/computational technique used to interpolate this data was Kernel Smoothing which, through non-parametric statistics, using the Gaussian kernel function, made it possible to estimate the entire area of the Northeast based on daily rainfall data from weather stations.

Using the "Kernel" software, it was possible to select the desired time interval and obtain the average and spatial standard deviation of the data for the period analyzed, as well as visualize the result obtained from interpolating the data, generating a map of average daily rainfall for the desired period.

Firstly, all the observations from 1094 to 1998 were interpolated, generating a spatial map that made it possible to visualize the average daily rainfall over the entire Northeast, revealing that the areas where rain is most abundant correspond to the north of the state of Maranhão and the east coast of Bahia, and the area with the least rainfall belongs to the semi-arid Northeast, especially on the border between Bahia and Pernambuco.

An investigation was also carried out into the accumulated annual rainfall for the years 1910 to 1998. The results obtained made it possible, by calculating 20% quantiles, to classify 1911 and 1993 as very dry years, and 1924, 1974 and 1985 as very rainy years.

To analyze whether there has been a change in the behavior of rainfall over time, rainfall data was interpolated into three periods of 30 years each. Based on the interpolations made, it was concluded that the amount of average daily rainfall has been gently decreasing over each period, but has maintained the same spatial characteristics, concentrating more on a large part of Maranhão and less on the semi-arid region.

With the help of the software, the monthly average daily rainfall was analyzed, making it possible to see that the seasons with the highest volume of rainfall are from December to February, and the driest months are September, October and November. The maps of monthly average daily rainfall made it possible to analyze the characteristics of the spatial distribution of rainfall in each month, revealing the movement of the areas of rain concentration over the Northeast.

We also analyzed the periods when rainfall occurs seasonally, corresponding to the following quarters: December, January and February; March, April and May; June, July and August. By interpolating the rainfall data for these periods and using the maps generated, it was possible to detect that from December to February, rainfall is concentrated in the western part of the Northeast, from March to May, rainfall predominates in the northern region, from June to August, rainfall is concentrated in a small part of the state of Maranhão and on the east coast, with the rest of the Northeast being quite dry.

The times when El Niño and La Niña occur were also investigated. Using the maps generated, it was possible to see that the average daily rainfall during the periods does not follow a pattern. By constructing the confidence interval for the difference in averages, it was seen that it cannot be said that during La Niña periods there is a predominance of rain in the Northeast, or during El Niño periods there is drought.

Based on the maps generated, it was possible to see that in some El Niño periods, rainfall did not undergo any apparent change, but most of the periods showed dry regions, with the period from February 1993 to September 1993 being the most punished by low rainfall. The La Niña periods, on the other hand, showed some dry regions, but most of them had areas of high rainfall concentration, with the period from September 1984 to June 1985 standing out.

The Pacific Decadal Oscillation was also analyzed in this work, revealing that there doesn't seem to be much difference between the average daily rainfall between the warm and cold phases.

This work represents the beginning of a series of studies on the behavior of rainfall over the Northeast of Brazil with a gigantic mass of data. This database could certainly not have been processed using conventional CPU technology. The "Kernel" software made it possible to do the interpolations from code written in C and CUDA, using the "Neumann" GPU cluster at UFRPE's Department of Statistics and Computer Science.

In the future, other studies could be carried out using the database analyzed. Other interpolators could be compared to see which can most accurately explain the behavior of rainfall over the Northeast. This study will probably help the relevant authorities make better use of the Northeast's climatological diversity.

References

ALI, A. Nonparametric spatial rainfall characterization using adaptive kernel estimator. **Journal of Geographic Information and decision Analysis**, v. 2, n. 2, p. 34-43, 1998.

ALVES, J. M. B.; REPELLI, C. A. A variabilidade pluviometrica no setor norte do nordeste e os eventos el nino-oscilacao sul (enos). **Revista Brasileira de Meteorologia**, v. 7, n. 2, p. 583-592, 1992.

ANDREOLI, R.; KAYANO, M. Multi-scale variability of the sea surface temperature in the tropical atlantic. **Journal of Geophysical Research**, v. 109, 2004.

ANDREOLI, R. V.; KAYANO, M. T. The relative importance of the tropical south atlantic and pacific east in the variability of precipitation in northeastern brazil. **Revista Brasileira de Meteorologia**, v. 22, n. 1, p. 63-74, 2007.

APAC. **Pernambuco Water and Climate Agency**. Available at: <http://www.apac.pe.gov.br/>.

BERNARDO, S. D. O.; MOLION, L. C. B. Comparison between observed precipitation totals and those estimated by cdc/ncep reanalyses for the northeast coast of brazil. **XII Congresso Brasileiro de Meteorología, Foz do Iguagu-PR**, 2002.

CARITAS-BRASILEIRA. **Agua de chuva - O segredo da convivencia com o Semi-arido brasileiro**. 2. ed. Sao Paulo, SP, Brazil: [s.n.], 2001.

CARNEIRO, E. O.; SANTOS, R. L. Spatial analysis applied to the determination of risk areas for some edemic diseases (kala-azar, dengue, diarrhea, STDs and tuberculosis), in the neighborhood of campo limpo, feira santana (ba). - sexually transmitted diseases and tuberculosis), in the neighborhood of campo limpo, feira de santana (ba). **sitientibus**, n. 28, p. 51-75, June 2003.

CHU, P.-S. Diagnostics studies of rainfall anomalies in northeast brazil. **American Meteorological Society**, v. 111, April 1983.

CHUNG, M. K. Heat kernel smoothing on unit sphere. **Department of Statistics, Biostatistics, and Medical Informatics. Waisman laboratory for brain imaging and behavior, University of Wisconsin**, p. 992-995, 2006.

CONTI, J. B. A questão climática do nordeste brasileiro e os processos de desertificacao. **Revista Brasileira de Climatologia**, v. 1, n. 1, 2005.

CORTES, R. X. A comparative study of additive non-parametric regression estimators: Performance in finite samples. **Monograph presented for the degree of Bachelor of Statistics - Federal University of Rio Grande do Sul - Institute of Mathematics - Department of Statistics**, 2004.

COSTA, C. C. L. da. **Meteorology**. July 2006. Available at: <http://pt.scribd.com/doc/66143648/Meteorologia>. Accessed on: 15/02/2012.

COSTA, J. de A. The el nino phenomenon and droughts in northeastern Brazil. **Paper presented for the degree of Licenciatura Plenae in Geography by the Federal University of Pernambuco**, 2009.

DUONG, T. ks: Kernel density estimation and kernel discriminant analysis for multivariate data in r. **Journal of Statistical Software**, v. 21, n. 7, october 2007.

EPANECHNIKOV, V. A. Non-parametric estimation of a multivariate probability density. theory of probability and its applications. v. 14, n. 1, p. 153-158, 1969.

FERREIRA, A. G. **Meteorología Pratica**. Sao Paulo, SP Brazil: [s.n.], 2006.

FERREIRA, A. G.; MELLO, N. G. da S. Main atmospheric systems acting over the northeastern region of brazil and the influence of the pacific and atlantic oceans on the region's climate. **Revista Brasileira de Climatologia,** v. 1, n. 1, p. 15 - 28, December 2005.

FERREIRA, M. R. P. Classical and nucleo discriminant analysis: Evaluations and some contributions regarding boosting and bootstrap methods. **Dissertation (master's degree) - Federal University of Pernambuco, CCEN, Statistics**, 2007.

FREIRE, J. L. M.; LIMA, J. R. A.; CAVALCANTI, E. P. Analysis of meteorological aspects over northeastern brazil in el nino and la nina years. **Revista Brasileira de Geografia Física**, v. 3, p. 429-444, September 2011.

FREITAS, A. C. V.; FRANCHITO, S. H.; RAO, V. B. Analysis of precipitation data from different sources over South America, with emphasis on Brazil. **CLIMEP - Climatologia e Estudos da Paisagem, Rios Claros, PR, Brazil,** v. 5, n. 1, June 2010.

FREITAS, A. R. de et al. Application of the kernel technique to cattle weighing data. Proceedings of **the IV National Symposium on Animal Improvement**, 2002.

GARCIA-PINTADO, J. et al. Rainfall estimation by rain gauge-radar combination: A concurrent multiplicative-additive approach. **Walter Resources Reasearch**, v. 45, p. 1-15, 2009.

GRIMM, A. M.; FERRAZ, S. E. T.; GOMES, J. Precipitation anomalies in southern brazil associated with el niño and la niña events. **American Meteorological Society**, v. 11, p. 2863-2880, November 1998.

HARTKAMP, A. D. et al. Interpolation techniques for climate variables. **CIMMYT Natural Resources Group**, 1999. Available at: <http://repository.cimmyt.org/xmlui/bitstream/handle/10883/988/67882.pdf?sequence=1>.

HARTMANN, D. L. **Global Physical Climatology**. Academic press. Seattle, Washington: Department of Atmospheric Sciences, University of Washington, 1994.

HASTENRATH, S. Prediction of northeast brazil rainfall anomalies. **American Meteorological Society**, v. 3, p. 893-904, August 1990.

HASTENRATH, S.; GREISCHAR, L. Circulation mechanisms related to northeast brazil rainfall anomalies. **JOURNAL OF GEOPHYSICAL RESEARCH**, v. 98, n. D3, p. 5093-5102, November 1993.

IBGE. **BRAZILIAN INSTITUTE OF GEOGRAPHY AND STATISTICS**. 2012. Available at: <http://www.ibge.gov.br/home/>.

JOU, P. H.; AKHOOND-ALI, A. M.; NAZEMOSADAT, M. J. Nonparametric kernel estimation of annual precipitation over iran. **Theoretical and Applied Climatology**, July 2012.

KALNAY, E. et al. The ncep/ncar 40-year reanalysis project. **Bulletin of the American Meteorogical Society**, p. 437-471, 1996.

KAWAMOTO, M. T. Analysis of spatial distribution techniques with point patterns and application to traffic accident data and dengue fever data from rio claro-sp. **Dissertation (master's degree) - Universidade Estadual Paulista - Instituto de Biociencias de Botucatu**, February 2012.

LI, M.; SHAO, Q. An improved statistical approach to merge satellite rainfall estimates and raingauge data. **Journal of Hydrology**, v. 385, p. 51-64, February 2010.

MANTUA, N. et al. A pacific interdecadal climate oscillation with impacts on salmon production. **Bulletin of the American Meteorological Society**, v. 78, p. 1069-1079, 1997.

MARENGO, J. A. **Characterization of the climate in the 20th century and scenarios for Brazil and South America for the 21st century derived from IPCC climate models**. [S.l.], 2007.

MATKAN, A. et al. A comparison between kriging, cokriging and geographically weighted regression models for estimating rainfall over north west of iran. **EMS Annual Meeting Abstracts**, v. 7, 2010.

MELLO, C. R. et al. Kriging and inverse of the square of the distance for interpolation of the parameters of the heavy rainfall equation. **Revista Brasileira de Ciencia do Solo**, v. 27, n. 5, p. 925-933, Oct. 2003.

MENDONCA, F.; DANNI-OLIVEIRA, I. M. **Climatologia: Nocoes Basicas e Climas do**

Brasil. Sao Paulo: [s.n.], 2007.

MENEGHETTI, G. T.; FERREIRA, N. J. Seasonal and interannual variability of precipitation in northeastern Brazil. **Anais XIV Simpósio Brasileiro de Sensoriamento Remoto, INPE**, p. 1685-1689, April 2009.

MOLION, L. C. B. Global warming, el ninos, sunspots, volcanoes and the Pacific decadal oscillation. **Revista Climanalise**, 2003.

MOLION, L. C. B.; BERNARDO, S. de oliveira. A review of rainfall dynamics in northeastern Brazil. **Revista Brasileira de Meteorologia**, v. 17, n. 1, p. 1-10, 2002.

NIKOLOVA, N.; VASSILEV, S. Mapping precipitation variability using different interpolation methods. **Proceedings of the Conference on Water Observation and Information System for Decision Support (BALWOIS)**, p. 25-29, May 2006.

NIMER, E. **Climatologia do Brasil**. 2. ed. Rio de Janeiro, RJ, Brazil: [s.n.], 1989.

NOBRE, C. A. et al. Aspects of the dynamic climatology of brazil. In: **Climanálise: Bulletin of Climate Monitoring and Analysis**. Brasília, DF, Brazil: INSTITUTO DE PESQUISAS ESPACIAIS - INPE and INSTITUTO NACIONAL DE METEOROLOGIA - INEMET, 1986.

OLIVEIRA, P. T. S. et al. Spatial variability of the rainfall erosive potential in the state of mato grosso do sul, brazil. **Engenharia Agrícola**, v. 32, n. 1, p. 69-79, Feb. 2012.

PARZEN, E. On estimation of a probability density function and mode. v. 33, p. 1065-1076, 1962.

PAULA, G. M. d. The el nino southern oscillation phenomenon and rainfall erosivity in santa maria - rs. **Dissertation presented to the Master's Course of the Postgraduate Program in Agricultural Engineering at the Federal University of Santa Maria (UFSM, RS), for the degree of Master in Agricultural Engineering**, 2009.

PIMENTEL, B. A. Kernel methods for interval data clustering. **Federal University of Pernambuco - Computer Center**, December 2010.

PINTO, C. C. de X. Profit diversity among Brazilian small businesses: the credit market as one of its possible determinants. **Dissertation (master's degree) - Pontifical Catholic University of Rio de Janeiro, Department of Economics,** May 2003.

RAO, V. B.; LIMA, M. C. de; FRANCHITO, S. H. Seasonal and interannual variations of rainfall over eastern northeast brazil. **Journal of Climate**, v. 6, p. 1754-1763, January 1993.

REIS, M. H. d. et al. Spatialization of precipitation data and evaluation of interpolators for agricultural drainage projects in the state of goias and federal district. **Anais XII Simpósio Brasileiro de Sensoriamento Remoto**, p. 229-236, 2005.

ROSENBLATT, M. Remarks on some nonparametric estimates of a density function. v. 27, n. 3, p. 832-837, September 1956.

SA, I. B.; SILVA, P. C. da. **Semiarido Brasileiro: Research, Development and Innovation**. 1. ed. Petrolina, PE: [s.n.], 2010.

SCHEID, S. **Introduction to Kernel Smoothing**. January 2004. Available at: <http://compdiag.molgen.mpg.de/docs/talk_05_01_04_stefanie.pdf>.

SCOTT, D. W. **Multivariate Density Estimation, Theory, Practice, and Visualization**. [s.n.], 1950. Available at: <http://www.stat.rice.edu/ scottdw/stat550/mde-92.pdf>.

SILVA, A. P. N. et al. Correlation between sea surface temperatures and the amount of rainfall during the rainy season in the northeast of the state of Pernambuco. **Revista Brasileira de Meteoorologia**, v. 26, n. 1, p. 149-156, June 2011.

SILVA, D. F. d.; GALVíNCIO, J. D. Study of the influence of the pacific decadal oscillation in northeastern brazil. **Revista Brasileira de Geografia Física**, v. 4, p. 665-676, May 2011. Available at: <http://www.ufpe.br/rbgfe/index.php/revista/article/view/142/198>. Accessed on: 06/03/2012.

SILVA, K. R. da et al. Spatial interpolation of precipitation in the state of Espírito Santo. **Floresta e Ambiente**, v. 18, n. 4, p. 417-427, December 2011.

SILVA, M. A. V. ao. **Meteorology and Climatology**. Digital version 2. Recife, Pernambuco, Brazil: [s.n.], 2006.

SOENARIO, I.; PLIEGER, M.; SLUITER, R. **Optimization of Rainfall Interpolation**. De Bilth, province of Utrecht, Netherlands, March 2010.

STECK, T. Methods for determining regularization for atmospheric retrieval problems. **Applied Optics**, v. 41, n. 9, p. 1788-1797, 2002.

SUDENE. **Superintendency for the Development of the Northeast**. Available at: <http://www.sudene.gov.br/>.

SUDENE. **SUPERINTENDENCE FOR THE DEVELOPMENT OF THE NORTHEAST**. 2012.
Available at: <http://www.sudene.gov.br/>.

TOMCZAK, M. Spatial interpolation and its uncertainty using automated anisotropic inverse distance weighting (idw) - cross-validation/jackknife approach. **Journal of Geographic Information and Decision Analysis**, v. 2, n. 2, p. 18-30, 1998.

TRENBERTH, K. E. The definition of el niño. **Bulletin of the American Meteorological Society**, v. 78, p. 2771-2777, 1997.

UPPALA, S. M. et al. The era-40 re-analysis. **Quarterly Journal of the Royal Meteorological Society**, v. 131, n. 612, p. 2961-3012, Oct 2005.

UVO, C.; BERNDTSSON, R. Regionalization and spatial properties of ceara state rainfall in northeast brazil. **JOURNAL OF GEOPHYSICAL RESEARCH**, v. 101, p. 4221-423, 1996.

UVO, C. B. et al. The relationships between tropical pacific and atlantic sst and northeast brazil monthly precipitation. **J. Climate**, v. 11, p. 551-562, 1998.

VERWORN, A.; HABERLANDT, U. Spatial interpolation of hourly rainfall - effect of additional information, variogram inference and storm properties. **Hydrology and Earth System Sciences**, v. 15, p. 569-584, 2011.

VIEIRA, L.; PICULLI, F. J. **Meteorologia e Climatologia Agricola-Class notes**. Cidade Gaúcha, PR, Brazil, 2009. Available at: <http://www.dea.uem.br/disciplinas/meteorologia/meteorologia8.pdf>.

XIONG, L.; GUO, S.; O'CONNOR, K. M. Smoothing the seasonal means of rainfall and runoff in the linear perturbation model (lpm) using the kernel estimator. **Journal of Hydrology**, v. 324, p. 266-282, 2006.

Printed by Books on Demand GmbH, Norderstedt / Germany